3 Texas STAAR Grade 3 Math Practice Tests

Full-Length Test Prep with Detailed Answer Explanations

Dr. A. Nazari

Published by View Math Education

ViewMath.com

3 Practice Tests to Get You Started!

Hey there, future math whiz! 

*This book has **3 full practice tests** to help you warm up for the real thing. Think of it like stretching before a big game — these tests will get your brain ready and show you what to expect!*

*Three tests is the **perfect start**!*

*Each one helps you feel **more ready**!*

*You'll be surprised how much you **already know**!*

Sharpen your pencil and let's get warmed up! 

Three practice tests is a great way to start. Take your time with each one, and you'll feel more confident every step of the way!

📖 How to Use This Book 📖

Your quick-start guide to 3 great practice tests!

📋 What's Inside This Book

- **3 Full-Length Practice Tests** — Each one covers all the Grade 3 math topics you need to know!

- **Answer Key with Explanations** — Find out why each answer is correct, not just what the answer is.

- **Reference Pages** — A math symbols chart and multiplication table you can peek at any time.

- **A Test Tracker** — Write down your scores and watch your confidence grow!

📅 A Simple 3-Test Plan

With just 3 tests, here's a great way to use them:

- **Test 1** — **The Warm-Up.** Take this test without a timer. Get comfortable with the question types. Don't worry about your score — just do your best!

- **Test 2** — **The Practice Round.** After reviewing Test 1, try this one with a timer (ask a grown-up!). Focus on the topics that were tricky last time.

- **Test 3** — **The Real Deal.** Treat this like the actual test: quiet room, timed, no peeking at answers. See how much you've improved!

● Multiple Choice

Pick the **one best answer** from choices A, B, C, or D. Not sure? Cross out the ones you know are wrong, then pick from what's left. That's a smart move!

✏️ Short Answer

Write your answer **and** show your work! Even if your final answer isn't right, showing your steps can earn you credit. Use scratch paper if you need more room.

66 After Each Test 99

Flip to the Answer Key and check your work. For every question you got wrong, **read the explanation carefully.** Then write the tricky topics on your Test Tracker page. If you need extra help, grab our **Grade 3 Math Study Guide!**

⭐ **Fun fact:** Three tests is all it takes to see real improvement! Most kids feel way more confident after just a few rounds of practice. ⭐

Find more at
ViewMath.com/TX-Grade3

 Tips for Test Day

Easy tricks to help you feel calm and do your best!

🌙 The Night Before

- ✅ **Sleep early** — *your brain learns while you sleep!*
- ✅ **Pack your supplies** — *pencils, eraser, scratch paper, all ready to go.*
- ✅ **Tell yourself:** *"I've been practicing. I'm going to do great!"*

👆 5 Simple Rules for Every Test

1. **Read the question twice.** *The first time to understand it. The second time to catch details.*
2. **Show your work.** *Write the steps down, even on scratch paper. It helps you think!*
3. **Skip the hard ones.** *Put a small star next to tricky questions and come back later. Answer the easy ones first!*
4. **Never leave a blank.** *For multiple choice, your best guess is better than no answer at all.*
5. **Check your work.** *Finished early? Go back and re-read your answers.*

✅ Smart Moves

- Take a deep breath before you begin
- Underline key words in the question
- Use drawings or number lines to help
- Cross out wrong answers first
- Double-check addition and subtraction

❌ Traps to Avoid

- Rushing and not reading carefully
- Picking the first answer that "looks right"
- Forgetting to carry or borrow numbers
- Skipping a question permanently
- Panicking when you see a tough problem

> ❝ *Remember, the very first practice test is the hardest — not because the questions are harder, but because everything is new! By Test 3, you'll feel like a pro. Trust me!* ❞

🧰 Get Ready to Practice 🧰

Pencils

Sharpened and ready!

Eraser

Everyone makes mistakes!

Scratch Paper

For working things out

A Calm Spot

Somewhere quiet to focus

A Grown-Up

To help set a timer

A Can-Do Attitude

You've totally got this!

✅ Allowed During Tests

- Pencils and erasers
- Blank scratch paper
- The **reference pages** in this book
- A ruler (for measurement questions)

❌ Not Allowed

- Calculators
- Phones, tablets, or computers
- Help from anyone else
- Your study guide (save it for after!)

👥 For Parents & Teachers

- With only 3 tests, **space them at least a week apart.** This gives time to review mistakes before trying the next one.
- Let your child take Test 1 untimed to build familiarity.
- After each test, go through the Answer Key together. Focus on **understanding the "why,"** not just the score.
- If a topic keeps tripping them up, review it in our **Grade 3 Math Study Guide** before the next practice test.
- Celebrate every bit of progress — even getting one more question right is a win!

X^1 Math Reference Sheet X^1

You may use this page during your practice tests!

Symbol	Name	What It Means	
$+$	Plus (Add)	Put numbers together.	$3 + 5 = 8$
$-$	Minus (Subtract)	Take away from a number.	$9 - 4 = 5$
$\times$	Times (Multiply)	Add equal groups.	$4 \times 3 = 12$
$\div$	Divide	Split into equal groups.	$12 \div 3 = 4$
$=$	Equals	Both sides are the same.	$2 + 3 = 5$
$>$	Greater Than	The left number is bigger.	$7 > 3$
$<$	Less Than	The left number is smaller.	$2 < 9$
$\frac{1}{2}$	Fraction Bar	Part of a whole.	$\frac{1}{2}$ means 1 out of 2 equal parts

📖 Key Math Words

- **Sum** — the answer when you add
- **Difference** — the answer when you subtract
- **Product** — the answer when you multiply
- **Quotient** — the answer when you divide
- **Factor** — a number you multiply
- **Array** — objects in rows and columns
- **Fraction** — a part of a whole
- **Numerator** — the top number in a fraction
- **Denominator** — the bottom number
- **Equation** — a math sentence with $=$
- **Estimate** — a smart guess, close to the real answer
- **Perimeter** — the distance around a shape
- **Area** — the space inside a shape
- **Rounding** — making a number simpler by going to the nearest ten or hundred

🔍 Word Problem Clue Words

- **Add** (+): in all, total, altogether, combined, sum, both, more

- **Subtract** (−): how many more, how many left, fewer, difference, remain

- **Multiply** (×): each, every, groups of, times, rows of, per

- **Divide** (÷): share equally, split, each group, how many groups, per

Find more at
ViewMath.com/TX-Grade3

▦ Multiplication Table ▦

×	1	2	3	4	5	6	7	8	9	10	11
1	1	2	3	4	5	6	7	8	9	10	11
2	2	4	6	8	10	12	14	16	18	20	22
3	3	6	9	12	15	18	21	24	27	30	33
4	4	8	12	16	20	24	28	32	36	40	44
5	5	10	15	20	25	30	35	40	45	50	55
6	6	12	18	24	30	36	42	48	54	60	66
7	7	14	21	28	35	42	49	56	63	70	77
8	8	16	24	32	40	48	56	64	72	80	88
9	9	18	27	36	45	54	63	72	81	90	99
10	10	20	30	40	50	60	70	80	90	100	110
11	11	22	33	44	55	66	77	88	99	110	121

💡 How to Use This Table

To find **4 × 7**:

1. Find **4** in the left column (blue).
2. Find **7** in the top row (blue).
3. Follow the row and column until they meet: the answer is **28**!

📈 My Confidence Tracker 📈

Record your scores below. You'll be amazed at your progress!

My name: _______________________________________

☑ Test	📅 Date	⭐ Score	🙂 How I Feel
1		/	
2		/	
3		/	

💡 My Quick Reflection

The easiest topic for me was:

____

The trickiest topic for me was:

____

One thing I got better at from Test 1 to Test 3:

____

Next time I want to try:

____

> *You just finished 3 practice tests — that's awesome! Compare your first score to your last. I bet you'll see real improvement. Ready for more? Check out our 5-test or 7-test books for even more practice!*

Find more at
ViewMath.com/TX-Grade3

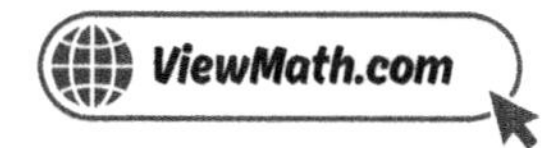

 Table of Contents

Here's what we'll explore together!

 *Let's learn and have fun!*

1

Practice Test 1

 30 Questions

 Before You Start 

✓ **Read each question carefully** before choosing your answer.

✓ **Show your work** on scratch paper when you need to.

✓ **Skip hard questions** and come back to them later.

✓ **Check your answers** when you're done.

✓ **Take your time** — there's no rush!

 You've Got This! 

Do your best and show what you know!

1. *Write* 5,600 *in expanded form.*

   Your Answer:

2. *Which number rounds to 500 when rounded to the nearest 100?*

   (A) 428

   (B) 449

   (C) 462

   (D) 551

3. *Which group contains only even numbers?*

   (A) 12, 34, 57

   (B) 20, 46, 88

   (C) 31, 50, 72

   (D) 14, 63, 90

4. *A school raised $8,000 last year and $2,000 this year. How much did the school raise in total?*

   (A) $6,000

   (B) $10,000

   (C) $82,000

   (D) $28,000

5. *What is the value of the digit 8 in 283,041?*

   (A) 80

   (B) 800

   (C) 8,000

   (D) 80,000

6. *What is* $198 + 345 + 267?$

   Your Answer:

Find more at
ViewMath.com/TX-Grade3

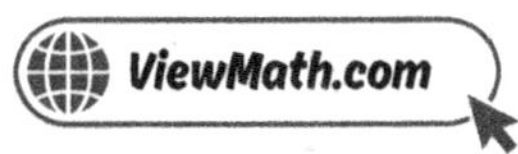

7. What is $6,789 + 3,456$?

Your Answer:

8. What is the missing digit? $8,\_\_46 - 3,289 = 4,757$

(A) 0

(B) 1

(C) 2

(D) 3

9. If $6 \times 9 = 54$, what is 9×6?

Your Answer:

10. What is $56 \div 8$?

(A) 6

(B) 7

(C) 8

(D) 48

11. Odd $\times$ odd $= ?$

(A) Always even

(B) Always odd

(C) Sometimes even, sometimes odd

(D) Always 0

12. What is 8×15?

(A) 80

(B) 100

(C) 115

(D) 120

Find more at
ViewMath.com/TX-Grade3

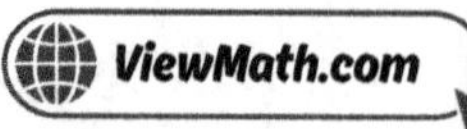

13. You eat $\frac{2}{6}$ of a cake. How many equal parts does the whole cake have?

Your Answer:

14. What whole number does $\frac{6}{3}$ equal?

(A) 1

(B) 2

(C) 3

(D) 6

15. You are at $\frac{3}{6}$ on a number line. You hop 2 more times by $\frac{1}{6}$. Where do you land?

Your Answer:

16. $\frac{3}{4} = \frac{?}{8}$. What is the missing numerator?

(A) 3

(B) 4

(C) 6

(D) 7

17. Tom ate $\frac{3}{8}$ of a pizza. Mia ate $\frac{5}{8}$ of the same pizza. Who ate more?

(A) Tom

(B) Mia

(C) They ate the same amount

(D) Cannot tell

18. You eat dinner at 6:30 in the evening. Is this A.M. or P.M.?

Your Answer:

Find more at
ViewMath.com/TX-Grade3

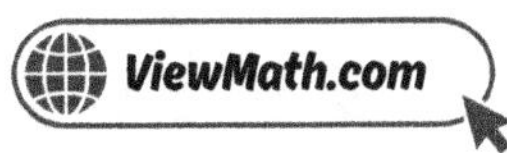

19. Emma started reading at 3:30 and stopped at 4:10. How long did she read?

    (A) 30 minutes

    (B) 40 minutes

    (C) 50 minutes

    (D) 1 hour and 10 minutes

20. 9,000 grams = how many kilograms?

    (A) 9 kg

    (B) 90 kg

    (C) 900 kg

    (D) 9,000 kg

21. A pitcher holds 8 liters of juice. The family drinks 3 liters. How much juice is left?

    (A) 3 L

    (B) 5 L

    (C) 8 L

    (D) 11 L

22. You have 3 quarters, 2 dimes, and 1 nickel. How much money do you have in cents?

    Your Answer

23. What is $2.50 + $1.25?

    (A) $3.25

    (B) $3.75

    (C) $4.75

    (D) $3.55

24. Jake has $25. He spends $9 on a book and $6 on supplies. How much does he have left?

    (A) $6

    (B) $10

    (C) $15

    (D) $16

25. A line plot shows pencil lengths. There are 3 X marks above $1\frac{1}{2}$ inches. What does this mean?

(A) 3 pencils are $1\frac{1}{2}$ inches long

(B) $1\frac{1}{2}$ pencils are 3 inches long

(C) The total length is $4\frac{1}{2}$ inches

(D) There are 3 pencils in all

26. Which statement about a rectangle is TRUE?

(A) All 4 sides are always equal.

(B) It has exactly 3 right angles.

(C) It has 4 right angles and 2 pairs of equal sides.

(D) It has no parallel sides.

27. How many faces does a hexagonal prism have?

(A) 6

(B) 7

(C) 8

(D) 10

28. A square has sides of 6 inches. What is the perimeter?

(A) 12 in

(B) 18 in

(C) 24 in

(D) 36 in

29. A pizza is cut into 2 equal slices. What fraction is each slice?

(A) $\frac{1}{2}$

(B) $\frac{1}{3}$

(C) $\frac{1}{4}$

(D) $\frac{2}{2}$

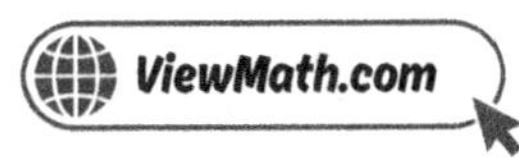

30. Carlos says a line segment goes on forever. Is he correct?

(A) Yes, line segments go on forever in both directions.

(B) Yes, line segments go on forever in one direction.

(C) No, a line segment has two endpoints and stops.

(D) No, a line segment has no endpoints.

 *End of Practice Test 1* 

Great job finishing the test!

☑ My Score

I got _____________ out of 30 questions right.

Check your answers in the **Answer Key** at the back of the book.

💡 *Review any questions you missed. That's how we learn!*

📊 Check Your Score Online!

Visit **ViewMath Academy** to enter your answers and see which topics you need to review. You can also explore lessons, take quizzes, track your scores, and save your progress!

viewmath.com/score/3.1.TX.01

Or go to viewmath.com/score and enter code: 3.1.TX.01

2

Practice Test 2

 30 Questions

✏️ Before You Start ✏️

- ✔ **Read each question carefully** before choosing your answer.
- ✔ **Show your work** on scratch paper when you need to.
- ✔ **Skip hard questions** and come back to them later.
- ✔ **Check your answers** when you're done.
- ✔ **Take your time** — there's no rush!

⭐ You've Got This! ⭐

Do your best and show what you know!

1. Write the expanded form of 9,205.

   Your Answer:

2. Maria rounded 438 to the nearest 10 and got 430. Is she correct?

   (A) No, it should be 440

   (B) Yes, 438 rounds to 430

   (C) No, it should be 400

   (D) No, it should be 450

3. List all the even numbers between 21 and 30.

   Your Answer:

4. How many hundreds are in 10,000?

   Your Answer:

5. Which is the correct expanded form of 720,056?

   (A) $72,000 + 56$

   (B) $700,000 + 20,000 + 50 + 6$

   (C) $700,000 + 2,000 + 56$

   (D) $700,000 + 20,000 + 56$

6. A library has 583 fiction books and 268 nonfiction books. How many books does the library have in all?

   Your Answer:

7. When adding $2{,}847 + 5{,}396$, which column do you start with?

   (A) Thousands   (B) Hundreds

   (C) Tens   (D) Ones

8. There were 5,000 tickets for a concert. If 3,812 tickets were sold, how many are still available?

   (A) 1,088   (B) 1,188

   (C) 1,288   (D) 2,188

9. If $7 \times 9 = 63$, what is 9×7?

   (A) 16   (B) 56

   (C) 63   (D) 79

10. In the division sentence $48 \div 6 = 8$, what is the divisor?

    (A) 48   (B) 6

    (C) 8   (D) 54

11. Looking at the diagonal of a multiplication table $(1 \times 1, 2 \times 2, 3 \times 3, \ldots)$, what are these numbers called?

    (A) Even numbers   (B) Odd numbers

    (C) Square numbers   (D) Prime numbers

12. What is 9×13?

    (A) 107   (B) 113

    (C) 117   (D) 127

Find more at
ViewMath.com/TX-Grade3

13. A pie is cut into 6 equal slices. 5 slices are left. What fraction of the pie is left?

(A) $\frac{1}{6}$

(B) $\frac{5}{6}$

(C) $\frac{6}{5}$

(D) $\frac{6}{6}$

14. Which fraction is closer to 1 on a number line divided into 6 parts?

(A) $\frac{2}{6}$

(B) $\frac{3}{6}$

(C) $\frac{4}{6}$

(D) $\frac{5}{6}$

15. Alex says $\frac{3}{4}$ is 3 copies of $\frac{1}{3}$. Is Alex correct?

(A) Yes

(B) No, $\frac{3}{4}$ is 3 copies of $\frac{1}{4}$

(C) No, $\frac{3}{4}$ is 4 copies of $\frac{1}{3}$

(D) No, $\frac{3}{4}$ is 3 copies of $\frac{3}{3}$

16. $\frac{1}{2} = \frac{?}{6}$. What is the missing number?

(A) 1

(B) 2

(C) 3

(D) 4

17. Is $\frac{5}{8}$ more or less than $\frac{1}{2}$?

(A) Less than $\frac{1}{2}$

(B) Equal to $\frac{1}{2}$

(C) More than $\frac{1}{2}$

(D) Cannot tell

18. You go to soccer practice at 4:00 in the afternoon. Is this A.M. or P.M.?

Your Answer:

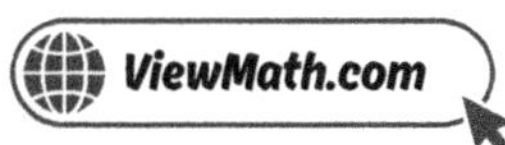

19. Jake starts his homework at 3:45 P.M. He works for 1 hour and 20 minutes. What time does he finish?

Your Answer:

20. A bag of rice has a mass of 5 kg. You use 2 kg for cooking. How much rice is left?

Your Answer:

21. Which container holds the LEAST liquid?

(A) A swimming pool

(B) A bathtub

(C) A bucket

(D) A teaspoon

22. Which group of coins equals 50 cents?

(A) 5 pennies

(B) 5 dimes

(C) 3 dimes

(D) 3 quarters

23. Lily has $6.00. She buys a sticker pack for $2.35 and a pen for $1.80. Does she have enough money left to buy a snack for $2.00?

(A) Yes, she has $2.85 left

(B) Yes, she has $2.00 left

(C) No, she only has $1.85 left

(D) No, she only has $1.65 left

24. You have $25. You buy a book for $12 and a snack for $3. How much do you have left, and is that enough to buy a $9 toy?

Your Answer:

Find more at
ViewMath.com/TX-Grade3

25. A line plot shows the lengths of 10 crayons. The data is shown below:

2 in.	$2\frac{1}{2}$ in.	3 in.	$3\frac{1}{2}$ in.	4 in.
2	3	4	1	0

How many crayons are shorter than 3 inches?

(A) 2

(B) 4

(C) 5

(D) 9

26. Every square is also a special kind of which shape?

(A) Triangle

(B) Pentagon

(C) Rectangle

(D) Trapezoid

27. How many vertices does a rectangular prism have?

(A) 4

(B) 6

(C) 8

(D) 12

28. A triangle has sides of 6 cm, 8 cm, and 10 cm. What is the perimeter?

Your Answer:

29. Which fraction is bigger: $\frac{1}{3}$ or $\frac{1}{6}$?

(A) $\frac{1}{6}$ is bigger

(B) $\frac{1}{3}$ is bigger

(C) They are the same size.

(D) You cannot tell.

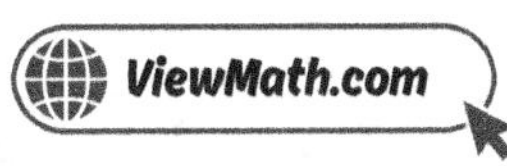

30. How many right angles does a square have?

Your Answer:

 End of Practice Test 2 

Great job finishing the test!

☑ My Score

I got _____________ out of 30 questions right.

*Check your answers in the **Answer Key** at the back of the book.*

💡 *Review any questions you missed. That's how we learn!*

📊 Check Your Score Online!

Visit **ViewMath Academy** to enter your answers and see which topics you need to review. You can also explore lessons, take quizzes, track your scores, and save your progress!

viewmath.com/score/3.1.TX.02

Or go to viewmath.com/score and enter code: 3.1.TX.02

3

Practice Test 3

 30 Questions

✏️ Before You Start ✏️

- ✔ **Read each question carefully** before choosing your answer.
- ✔ **Show your work** on scratch paper when you need to.
- ✔ **Skip hard questions** and come back to them later.
- ✔ **Check your answers** when you're done.
- ✔ **Take your time** — there's no rush!

★ You've Got This! ★

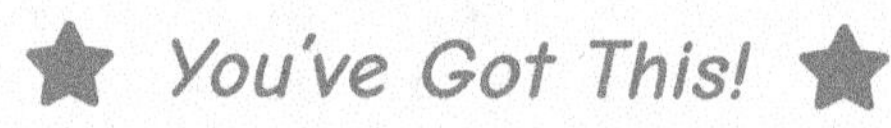 Do your best and show what you know!

1. *Each place value is how many times bigger than the place to its right?*

   (A) 2 times
   (B) 5 times
   (C) 10 times
   (D) 100 times

2. *When you round 549 to the nearest 10 and to the nearest 100, which gives the larger answer?*

   (A) Rounded to the nearest 10
   (B) Rounded to the nearest 100
   (C) They are the same
   (D) Cannot tell without rounding

3. *If you add two odd numbers, the result is always:*

   (A) Odd
   (B) Even
   (C) Greater than 10
   (D) An odd number less than 20

4. *What number is 1 more than 9,999?*

   (A) 9,000
   (B) 9,990
   (C) 10,000
   (D) 10,001

5. *What is the smallest 6-digit number?*

   Your Answer

6. *Which addition problem requires regrouping?*

   (A) $135 + 243$
   (B) $428 + 351$
   (C) $367 + 485$
   (D) $512 + 136$

Find more at
ViewMath.com/TX-Grade3

7. What is $2{,}103 + 5{,}642$?

(A) 7,745

(B) 7,645

(C) 7,755

(D) 8,745

8. A store had 6,250 items. They sold 3,487 items. How many items are left?

(A) 2,673

(B) 2,763

(C) 2,773

(D) 3,763

9. Which property is used here? $8 \times 7 = 8 \times (5 + 2) = (8 \times 5) + (8 \times 2)$

(A) Commutative property

(B) Associative property

(C) Distributive property

(D) Zero property

10. Carlos starts with 40 stickers. He gives away 5 stickers at a time. How many times can he give away stickers?

(A) 7

(B) 8

(C) 9

(D) 35

11. The rule for a pattern is "multiply by 3." If the first number is 2, what are the next two numbers?

(A) $5, 8$

(B) $4, 8$

(C) $6, 12$

(D) $6, 18$

12. Mom buys 6 bags of apples. Each bag has 18 apples. How many apples in all?

Your Answer:

Find more at
ViewMath.com/TX-Grade3

13. A shape is divided into parts, but the parts are NOT equal sizes. Can you write a fraction for the shaded part?

(A) Yes, just count the shaded parts

(B) Yes, fractions work with any parts

(C) No, fractions only work when parts are equal

(D) No, you need to shade all the parts first

14. A number line is divided into 4 equal parts. Which fraction is at the halfway point between 0 and 1?

(A) $\frac{1}{4}$

(B) $\frac{2}{4}$

(C) $\frac{3}{4}$

(D) $\frac{4}{4}$

15. Count by fourths: $\frac{1}{4}$, $\frac{2}{4}$, ___, $\frac{4}{4}$. What fraction is missing?

Your Answer:

16. $\frac{4}{8}$ simplified is:

(A) $\frac{1}{4}$

(B) $\frac{2}{4}$

(C) $\frac{1}{2}$

(D) $\frac{2}{8}$

17. When two fractions have the same denominator, how do you compare them?

(A) The one with the bigger denominator is larger

(B) The one with the bigger numerator is larger

(C) They are always equal

(D) You cannot compare them

18. A clock shows the short hand past the 10 and the long hand between the 2 and 3, on the second tick mark past the 2. What time does the clock show?

(A) 10:10

(B) 10:12

(C) 2:50

(D) 10:15

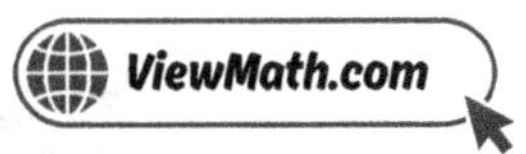

19. A library is open from 10:00 A.M. to 3:00 P.M. How long is it open?

(A) 3 hours

(B) 4 hours

(C) 5 hours

(D) 7 hours

20. A box of cereal has a mass of 500 g. You buy 2 boxes. What is the total mass?

(A) 502 g

(B) 700 g

(C) 1,000 g

(D) 5,000 g

21. A pot holds 6 liters when full. You pour in 2 liters first, then 3 more liters. How many more liters are needed to fill the pot?

Your Answer:

22. How much is a quarter worth?

(A) 1 cent

(B) 5 cents

(C) 10 cents

(D) 25 cents

23. What is $3.45 + $2.80?

(A) $5.25

(B) $6.25

(C) $5.65

(D) $6.15

24. Which is an example of EARNING money?

(A) Finding a $5 bill on the ground

(B) Getting paid for doing chores

(C) Borrowing money from a friend

(D) Spending money at a store

Find more at
ViewMath.com/TX-Grade3

---

25. A line plot shows the lengths of nails in inches. The data points are: 1, 1, $1\frac{1}{2}$, 2, 2, 2, $2\frac{1}{2}$, $2\frac{1}{2}$. How many X marks should be placed above 2 inches?

Your Answer:

26. Mia says, "All rectangles are squares." Is she correct?

(A) Yes, because both have 4 sides.

(B) Yes, because both have 4 right angles.

(C) No, because a rectangle does not always have 4 equal sides.

(D) No, because a rectangle has no right angles.

27. A pyramid has a hexagonal base (6 sides). How many faces, edges, and vertices does it have?

Your Answer:

28. Which unit would you use for perimeter?

(A) Square centimeters

(B) Centimeters

(C) Cubic centimeters

(D) Square inches

29. A square is partitioned into 4 equal parts. All 4 parts are shaded. What fraction is shaded?

(A) $\frac{1}{4}$

(B) $\frac{3}{4}$

(C) $\frac{4}{4}$

(D) $\frac{4}{1}$

30. How many right angles can you see in the capital letter "L"?

(A) 0

(B) 1

(C) 2

(D) 4

Find more at
ViewMath.com/TX-Grade3

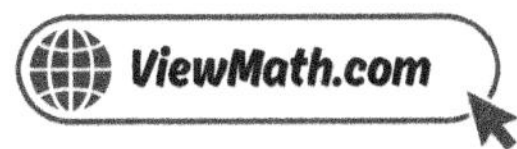

 End of Practice Test 3 

Great job finishing the test!

✅ My Score

I got ___________ out of 30 questions right.

*Check your answers in the **Answer Key** at the back of the book.*

💡 *Review any questions you missed. That's how we learn!*

📊 Check Your Score Online!

Visit **ViewMath Academy** to enter your answers and see which topics you need to review. You can also explore lessons, take quizzes, track your scores, and save your progress!

viewmath.com/score/3.1.TX.03

Or go to viewmath.com/score and enter code: 3.1.TX.03

Answer Key & Explanations

★ Check Your Answers! ★

First try each test on your own, then look here to check.

Read the explanations to learn from any mistakes ★

☑ Practice Test 1 — Answer Key

| 1 | $5,000 + 600$ | 2 | C | 3 | B | 4 | B | 5 | D | 6 | 810 | 7 | 10,245 | 8 | A |

9 · 54 | 10 · B | 11 · B | 12 · D | 13 · 6 | 14 · B | 15 · $\frac{5}{6}$ | 16 · C | 17 · B | 18 · P.M.

19 · B | 20 · A | 21 · B | 22 · 100 cents (or $1.00) | 23 · B | 24 · B | 25 · A | 26 · C

27 · C | 28 · C | 29 · A | 30 · C

💡 Time to Learn! 💡

Go through the explanations below, **especially for the questions you missed.**

Understanding why each answer is correct makes you a stronger math thinker!

👍 **Tip:** Circle any questions you got wrong, then read their explanation carefully.

📖 Practice Test 1 — Detailed Explanations

1. $5,600 = 5,000 + 600 + 0 + 0.$ The tens and ones are both 0.

Find more at
ViewMath.com/TX-Grade3
👉

2 *462 has tens digit $6 \geq 5$, so it rounds up to 500. 428 and 449 round to 400. 551 rounds to 600.*

3 *20 (ends in 0), 46 (ends in 6), 88 (ends in 8) are all even. The other groups each contain at least one odd number.*

4 *$8,000 + 2,000 = 10,000$.*

5 *In 283,041, the digit 8 is in the ten-thousands place, so its value is 80,000.*

6 *First add $198 + 345 = 543$. Ones: $8 + 5 = 13$, carry 1. Tens: $9 + 4 + 1 = 14$, carry 1. Hundreds: $1 + 3 + 1 = 5$. Then $543 + 267 = 810$. Ones: $3 + 7 = 10$, carry 1. Tens: $4 + 6 + 1 = 11$, carry 1. Hundreds: $5 + 2 + 1 = 8$.*

7 *Ones: $9 + 6 = 15$, carry 1. Tens: $8 + 5 + 1 = 14$, carry 1. Hundreds: $7 + 4 + 1 = 12$, carry 1. Thousands: $6 + 3 + 1 = 10$. The sum is $10,245$ — a 5-digit number! Two 4-digit numbers can add up to more than 9,999.*

8 *Add the difference and the subtracted number. $4,757 + 3,289 = 8,046$. The missing digit is 0.*

9 *By the commutative property, swapping the factors does not change the product.*

10 *$56 \div 8 = 7$ because $8 \times 7 = 56$.*

11 *Odd $\times$ odd always gives an odd product. Example: $3 \times 5 = 15$, $7 \times 9 = 63$.*

12 *$(8 \times 10) + (8 \times 5) = 80 + 40 = 120$.*

13 *The denominator tells the total equal parts. The cake is cut into 6 equal parts.*

14 *$\frac{3}{3} = 1$, so $\frac{6}{3} = 2$.*

Find more at
ViewMath.com/TX-Grade3

15 $\frac{3}{6} + \frac{1}{6} + \frac{1}{6} = \frac{5}{6}$.

16   *The denominator went from 4 to 8 (multiplied by 2). So multiply the numerator by 2 too: $3 \times 2 = 6$.*

17   *Same denominator. $5 > 3$, so $\frac{5}{8} > \frac{3}{8}$. Mia ate more.*

18   *Evening is after noon, so 6:30 in the evening is 6:30 P.M.*

19   *From 3:30 to 4:00 is 30 minutes. From 4:00 to 4:10 is 10 minutes. Total: $30 + 10 = 40$ minutes.*

20   *Divide by 1,000: $9,000 \div 1,000 = 9$ kg.*

21 $8 - 3 = 5$ *L.*

22 $3 \times 25 = 75$. *Then $2 \times 10 = 20$. Then $1 \times 5 = 5$. Total: $75 + 20 + 5 = 100$ cents.*

23   *Line up the decimals and add: $\$2.50 + \$1.25 = \$3.75$.*

24 $9 + 6 = 15$. $25 - 15 = 10$. *Jake has $10 left.*

25   *Each X mark represents one pencil. 3 X marks above $1\frac{1}{2}$ means 3 pencils measured $1\frac{1}{2}$ inches.*

26   *A rectangle has 4 right angles and 2 pairs of equal sides (opposite sides are equal). Not all 4 sides have to be equal.*

27   *A prism with an n-sided base has $n + 2$ faces. Hexagonal base: $6 + 2 = 8$ faces (2 hexagons + 6 rectangles).*

28   *A square has 4 equal sides, so perimeter $= 4 \times 6 = 24$ in.*

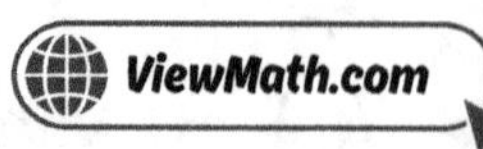

29 When a whole is split into 2 equal parts, each part is $\frac{1}{2}$ (one-half).

30 A line segment has two endpoints, so it does NOT go on forever. Lines and rays go on forever.

✅ Practice Test 2 — Answer Key

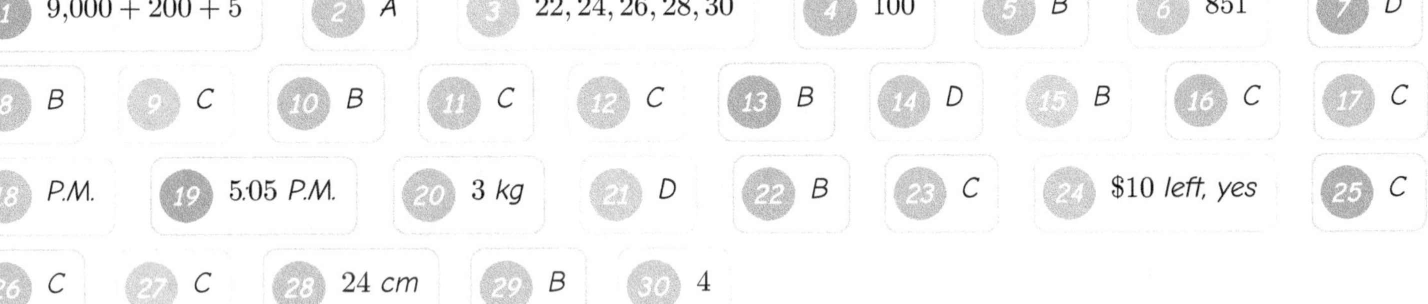

1 $9,000 + 200 + 5$	**2** A	**3** $22, 24, 26, 28, 30$
4 100	**5** B	**6** 851	**7** D
8 B	**9** C	**10** B	**11** C	**12** C	**13** B	**14** D	**15** B	**16** C	**17** C
18 P.M.	**19** 5:05 P.M.	**20** 3 kg	**21** D	**22** B	**23** C	**24** $10 *left, yes*	**25** C
26 C	**27** C	**28** 24 cm	**29** B	**30** 4

💡 Time to Learn! 💡

Go through the explanations below, **especially for the questions you missed**.

Understanding why each answer is correct makes you a stronger math thinker!

👍 **Tip:** Circle any questions you got wrong, then read their explanation carefully.

📖 Practice Test 2 — Detailed Explanations

1 $9,205 = 9,000 + 200 + 0 + 5$. The tens digit is 0, so there is no tens term.

2 The ones digit is 8. Since $8 \geq 5$, we round up. 438 rounds to 440, not 430.

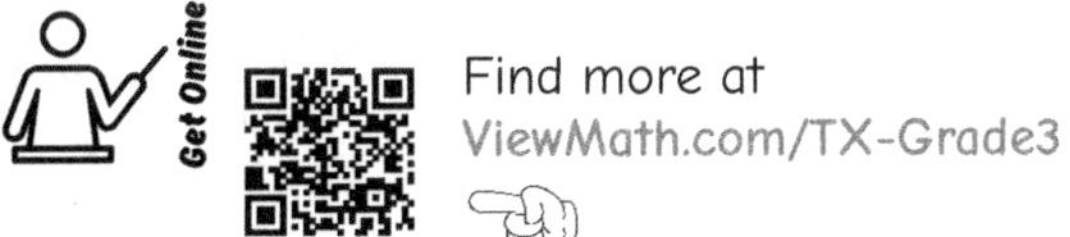

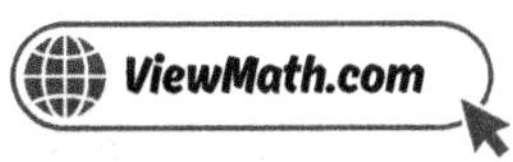

3 The even numbers between 21 and 30 are $22, 24, 26, 28, 30$. They all end in an even digit.

4 $10,000 \div 100 = 100$. There are 100 hundreds in 10,000.

5 $720,056 = 700,000 + 20,000 + 0 + 0 + 50 + 6$. The thousands and hundreds places are both zero.

6 $583 + 268 = 851$. Ones: $3 + 8 = 11$, carry 1. Tens: $8 + 6 + 1 = 15$, carry 1. Hundreds: $5 + 2 + 1 = 8$.

7 Always start adding from the ones column (the rightmost column) and work your way to the left.

8 $5,000 - 3,812 = 1,188$. Borrow through zeros: $10 - 2 = 8$, $9 - 1 = 8$, $9 - 8 = 1$, $4 - 3 = 1$.

9 By the commutative property, $9 \times 7 = 7 \times 9 = 63$.

10 The divisor is the number you divide by. In $48 \div 6 = 8$, the divisor is 6.

11 Numbers like $1, 4, 9, 16, 25, 36, \ldots$ where a number is multiplied by itself are called square numbers.

12 $(9 \times 10) + (9 \times 3) = 90 + 27 = 117$.

13 5 slices out of 6 equal slices $= \frac{5}{6}$.

14 $\frac{5}{6}$ is only $\frac{1}{6}$ away from 1, closer than all the other choices.

15 $\frac{3}{4}$ means 3 copies of $\frac{1}{4}$, not $\frac{1}{3}$. The denominator tells the size of each piece.

16 The denominator went from 2 to 6 (multiplied by 3). Multiply the numerator too: $1 \times 3 = 3$. So $\frac{1}{2} = \frac{3}{6}$.

Find more at
ViewMath.com/TX-Grade3

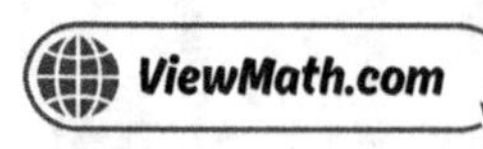

17 $\frac{1}{2} = \frac{4}{8}$. Since $\frac{5}{8} > \frac{4}{8}$, it is more than $\frac{1}{2}$.

18 Afternoon is after noon, so 4:00 in the afternoon is 4:00 P.M.

19 $3:45 + 1\ hour = 4:45$. Then $4:45 + 20\ minutes = 5:05$ P.M.

20 $5 - 2 = 3\ kg$.

21 A teaspoon holds much less than 1 liter, which is far less than a bucket, bathtub, or pool.

22 $5\ dimes = 5 \times 10 = 50\ cents$.

23 Total spent: $\$2.35 + \$1.80 = \$4.15$. Money left: $\$6.00 - \$4.15 = \$1.85$. Since $\$1.85 < \$2.00$, she does not have enough.

24 $12 + 3 = 15$. $25 - 15 = 10$. A \$9 toy costs less than \$10, so yes, there is enough.

25 Crayons shorter than 3 inches: 2 in. (2) $+ 2\frac{1}{2}$ in. (3) $= 5$ crayons.

26 A square has 4 right angles just like a rectangle, so every square is a special kind of rectangle. It also has 4 equal sides, making it a special rhombus too.

27 A rectangular prism has 8 vertices, 12 edges, and 6 faces. The formula for a prism with n-sided base is $2n$ vertices: $2 \times 4 = 8$.

28 Add all side lengths: $6 + 8 + 10 = 24\ cm$.

29 When the whole is the same size, fewer parts means bigger pieces. $\frac{1}{3}$ means 3 parts, $\frac{1}{6}$ means 6 parts. $\frac{1}{3} > \frac{1}{6}$.

Find more at
ViewMath.com/TX-Grade3

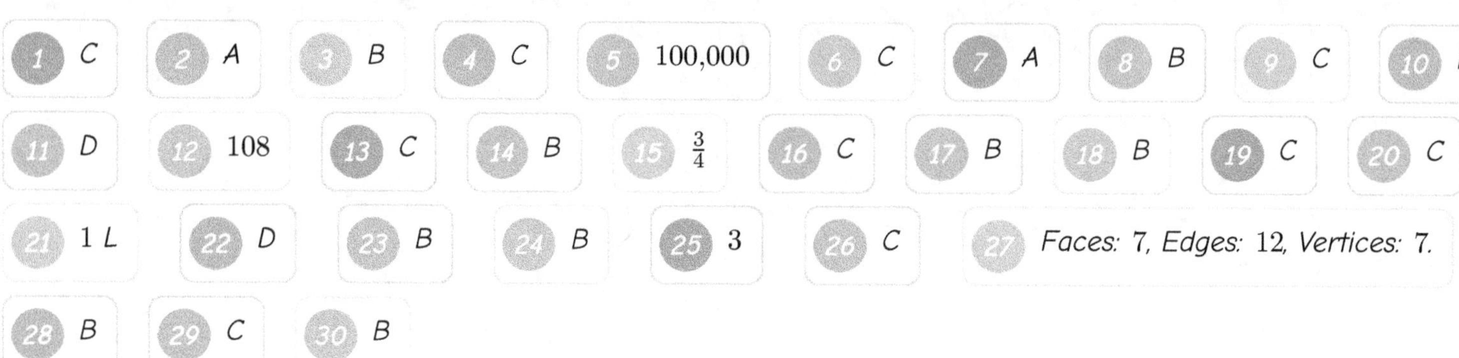

30 A square has 4 corners, and each one is a right angle.

☑ Practice Test 3 — Answer Key

| 1 | C | 2 | A | 3 | B | 4 | C | 5 | 100,000 | 6 | C | 7 | A | 8 | B | 9 | C | 10 | B |

| 11 | D | 12 | 108 | 13 | C | 14 | B | 15 | $\frac{3}{4}$ | 16 | C | 17 | B | 18 | B | 19 | C | 20 | C |

| 21 | 1 L | 22 | D | 23 | B | 24 | B | 25 | 3 | 26 | C | 27 | Faces: 7, Edges: 12, Vertices: 7. |

| 28 | B | 29 | C | 30 | B |

💡 Time to Learn! 💡

Go through the explanations below, **especially for the questions you missed.**

Understanding why each answer is correct makes you a stronger math thinker!

👍 **Tip:** Circle any questions you got wrong, then read their explanation carefully.

📖 Practice Test 3 — Detailed Explanations

1 Each place is 10 times bigger than the place to its right: ones → tens → hundreds → thousands.

2 549 rounded to the nearest 10: ones digit is $9 \geq 5$, so 550. Rounded to the nearest 100: tens digit is $4 < 5$, so 500. $550 > 500$.

3 Odd + Odd = Even. For example, $3 + 5 = 8$ (even) and $7 + 9 = 16$ (even).

Find more at
ViewMath.com/TX-Grade3

4 $9,999 + 1 = 10,000$. *This is the first 5-digit number.*

5   *The smallest 6-digit number is* $100,000$. *Any number below it (like 99,999) has only 5 digits.*

6   *In* $367 + 485$, *the ones column is* $7 + 5 = 12$, *which is 10 or more, so you must regroup. The other problems do not need regrouping.*

7   *No regrouping needed. Ones:* $3 + 2 = 5$. *Tens:* $0 + 4 = 4$. *Hundreds:* $1 + 6 = 7$. *Thousands:* $2 + 5 = 7$. *The sum is 7,745.*

8 $6,250 - 3,487 = 2,763$. *Borrow as needed in each column: ones* $10 - 7 = 3$, *tens* $14 - 8 = 6$, *hundreds* $1 - 4$, *borrow again:* $11 - 4 = 7$, *thousands* $5 - 3 = 2$.

9   *Breaking apart a factor and multiplying each part is the distributive property.*

10 $40 \div 5 = 8$. *He can give away stickers 8 times. This is like subtracting 5 from 40 repeatedly until reaching* 0, *which takes 8 steps.*

11 $2 \times 3 = 6$, *then* $6 \times 3 = 18$. *The next two numbers are* $6, 18$.

12 $6 \times 18 = (6 \times 10) + (6 \times 8) = 60 + 48 = 108$ *apples.*

13   *Fractions only work when the parts are* **equal**. *If the pieces are different sizes, you cannot write a fraction.*

14 $\frac{2}{4}$ *is at the second of 4 marks, which is the middle of the line from 0 to 1.*

15   *Counting by* $\frac{1}{4}$: $\frac{1}{4}, \frac{2}{4}, \frac{3}{4}, \frac{4}{4}$.

16   *Divide top and bottom by 4:* $\frac{4 \div 4}{8 \div 4} = \frac{1}{2}$.

Find more at
ViewMath.com/TX-Grade3

17 Same denominator means same-size pieces. More pieces (bigger numerator) = bigger fraction.

18 The short hand past 10 means the hour is 10. The long hand on 2 means 10 minutes. Two more tick marks gives $10 + 2 = 12$ minutes. The time is 10:12.

19 From 10:00 A.M. to 3:00 P.M.: count $10 \rightarrow 11 \rightarrow 12 \rightarrow 1 \rightarrow 2 \rightarrow 3$, which is 5 hours.

20 $500 + 500 = 1,000$ g (which is the same as 1 kg).

21 You poured in $2 + 3 = 5$ L. The pot holds 6 L, so $6 - 5 = 1$ more liter is needed.

22 A quarter is worth 25 cents.

23 Cents: $45 + 80 = 125$ cents $= 1$ dollar and 25 cents. Dollars: $3 + 2 + 1 = 6$. Answer: $6.25.

24 Earning money means working for it. Getting paid for doing chores is an example of earning money.

25 Count how many times 2 appears in the data: 2, 2, 2. That's 3 times.

26 A square must have 4 equal sides AND 4 right angles. A rectangle has 4 right angles but does not always have 4 equal sides (only when it is a square).

27 Pyramid with n-sided base: faces $= n + 1 = 7$, edges $= 2n = 12$, vertices $= n + 1 = 7$.

28 Perimeter is a distance, so it is measured in regular length units like centimeters or inches. Square units are for area.

29 All 4 out of 4 parts are shaded, so $\frac{4}{4}$ is shaded. $\frac{4}{4} = 1$ whole.

30 The letter "L" has one corner where the two lines meet at 90°, forming 1 right angle.

Great job checking your work!

Keep practicing and you'll be a math star!

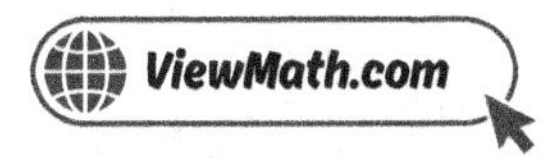

www.ingramcontent.com/pod-product-compliance
Lightning Source LLC
Chambersburg PA
CBHW081202130726
47996CB00009B/3224